FOCUS ON ARTIFICIAL INTELLIGENCE

ARTIFICIAL INTELLIGENCE AND DAILY LIFE

by Kari Cornell

BrightPoint Press

San Diego, CA

BrightPoint Press

an imprint of ReferencePoint Press, Inc.
Printed in the United States

For more information, contact:
BrightPoint Press
PO Box 27779
San Diego, CA 92198
www.BrightPointPress.com

LIBRARY OF CONGRESS CATALOGING-IN-PUBLICATION DATA

Names: Cornell, Kari A., author.
Title: Artificial intelligence and daily life / by Kari Cornell.
Description: San Diego, CA: BrightPoint Press, [2025] | Series: Focus on artificial intelligence | Includes bibliographical references and index. | Audience: Grades 7-9
Identifiers: LCCN 2024001081 (print) | LCCN 2024001082 (eBook) | ISBN 9781678209469 (hardcover) | ISBN 9781678209476 (eBook)
Subjects: LCSH: Artificial intelligence--Juvenile literature.
Classification: LCC Q335.4.C67 2025 (print) | LCC Q335.4 (eBook) | DDC 006.3--dc23/eng/20240201
LC record available at https://lccn.loc.gov/2024001081
LC eBook record available at https://lccn.loc.gov/2024001082

CONTENTS

AT A GLANCE 4

INTRODUCTION 6
AN AI WORLD

CHAPTER ONE 12
AI AND HOW IT WORKS

CHAPTER TWO 24
AI AT HOME

CHAPTER THREE 36
AI IN ENTERTAINMENT

CHAPTER FOUR 46
AI AT SCHOOL

Glossary 58
Source Notes 59
For Further Research 60
Index 62
Image Credits 63
About the Author 64

AT A GLANCE

- Artificial intelligence (AI) systems search through data for patterns and connections.
- Machine learning AI systems use patterns and connections in data to improve their ability to do tasks.
- AI has grown quickly because of the huge amount of data that is now available. Powerful computers are needed to process this data.
- AI systems are changing daily life at home, in entertainment, and in school.
- The AI system ChatGPT responds to prompts. It creates its answers one word at a time. ChatGPT decides what the most likely next word is based on the data it was trained on.

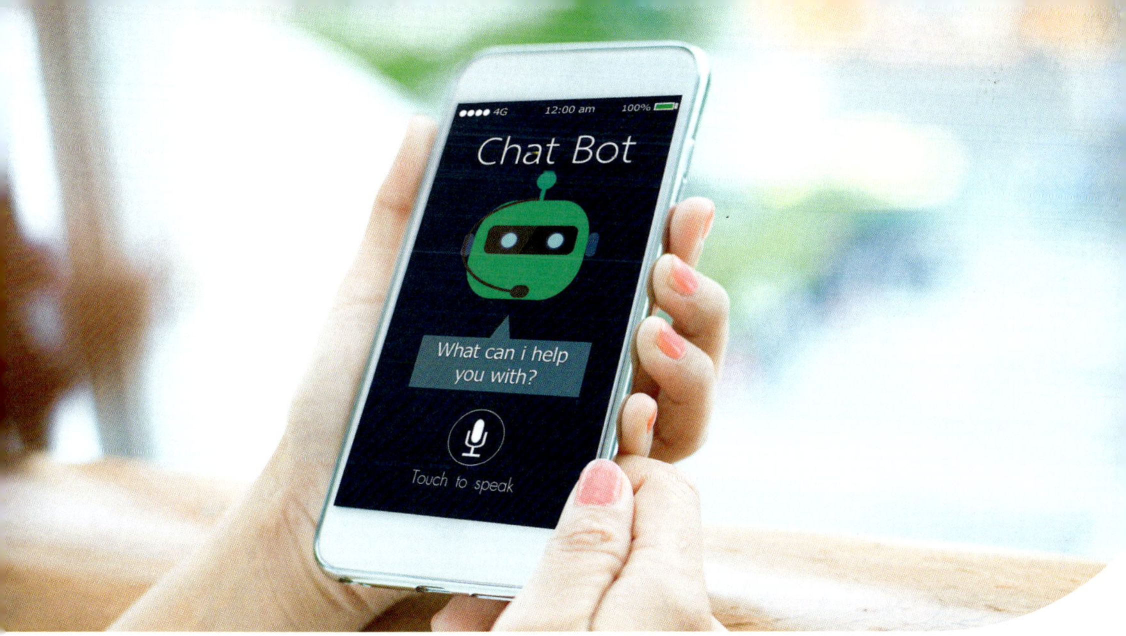

- The internet of things (IoT) refers to devices that can connect to the internet and one another. Some IoT devices used in people's homes include AI.

- In video games, computer-controlled characters can use AI to behave more like human players.

- A type of AI called deep learning can find patterns in the movies and shows that viewers watch. It can create lists of other programs viewers might like.

- AI in schools can help students learn and practice concepts. It can help teachers personalize lessons to each student's needs.

AN AI WORLD

It's an average day in the year 2040 for Grace. The 16-year-old gets out of bed and puts in her contact lenses. They help her see. But they also include her artificial intelligence (AI) assistant. Grace asks, "What is my schedule today?" When she looks at the wall, her lenses show her calendar. Her appointments seem to float on the wall in front of her. She has an AP Bio test at 10:00 a.m. Lacrosse practice is at 3:30 p.m.

Headsets, glasses, and even contact lenses can project information into a person's field of vision.

Search...

Grace gets dressed. Much of what she wears incorporates AI technology. Her earrings monitor her heart rate and body

Self-driving cars use AI to safely navigate roads.

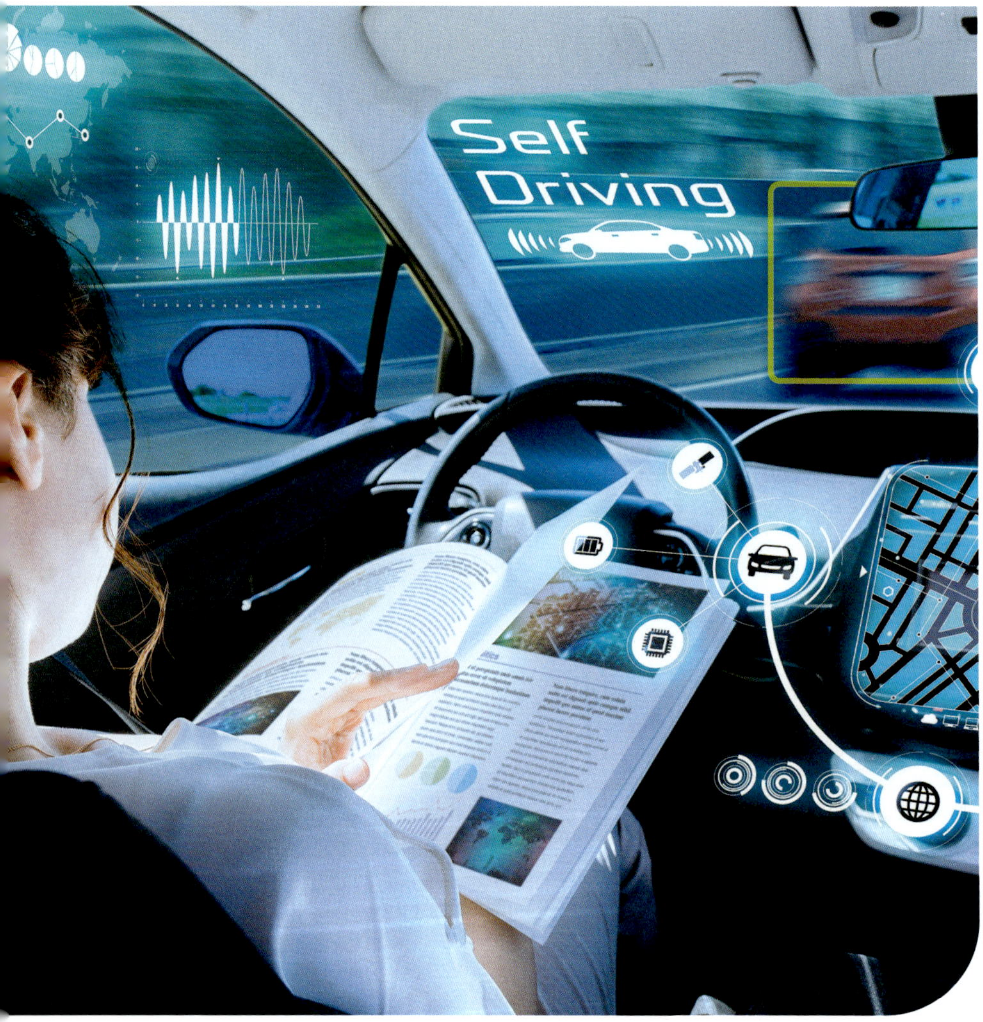

temperature. Her shoes convert heat and movement into energy. As she walks, they charge the tech she wears.

Grace goes downstairs. Ella has made her breakfast. Ella is the family robot. Grace eats quickly and leaves her dishes for Ella. An on-demand car is waiting to take Grace to school. Most new cars are self-driving, so few people need to drive. Grace reclines her seat. The car's computer knows her schedule. It sees that she has an AP Bio test. It displays her notes so she can study on the way to school.

WHAT IS ARTIFICIAL INTELLIGENCE?

In Grace's world, AI plays a role in nearly everything she does. Her story is fictional.

But it is based on experts' predictions about the future.

AI is the ability of a computer to do tasks usually done by humans. For some problems, AI can work more quickly than the human mind. AI gathers huge amounts of data. It studies the data for patterns and connections. It forms conclusions and learns from mistakes. AI can self-correct. It can make smart decisions. In the early 2020s, AI made big leaps forward. The technology has more uses than ever before. It is changing people's lives in many different ways.

Rooms full of powerful computers are behind today's cutting-edge AI technology.

CHAPTER ONE

AI AND HOW IT WORKS

Every day, people's lives are affected by AI. According to a 2022 Pew Research survey, 44 percent of Americans think they do not use AI regularly. But AI is everywhere. It's behind face recognition systems on smartphones. AI also suggests more movies after viewers watch something on Netflix. AI shows personalized ads after a user has searched for something in online

Unlocking smartphones using facial recognition is one place where AI is already in use today.

Machine learning technology can sort through vast amounts of data very quickly to make itself smarter.

stores. And AI creates the social media feeds that users see each day.

HOW DOES AI WORK?

AI is designed to act like the human mind. But AI does not have human experiences. These are what give humans knowledge. Knowledge helps humans solve problems and find answers. To make up for this lack of real-world experience, AI gathers huge amounts of data.

AI uses an algorithm. An algorithm is a set of rules used to solve a problem. It searches the data for patterns and connections. Patterns in the data become the building blocks of AI knowledge. Some AI systems gather data. Then they study the patterns to create rules. Finally, they apply those rules to process new data. This is known as **machine learning (ML)**.

Computer technology is getting better each year. As it improves, so does AI. AI has grown quickly. This is because of the enormous amount of data now available online. Huge amounts of storage are needed to hold all that data. Powerful computers are needed to process it.

The quality of the data has also improved. AI systems can access books and online articles. They can also search images and videos. All this data helps AI see how things are connected.

With new data and technology, AI will continue to improve. Bill Gates is the founder of the technology company Microsoft. In 2023, he wrote, “Developing AI . . . has been the great dream of the computing industry. For decades, the question was when computers would be better than humans at something other than making calculations. Now, with the arrival of machine learning and large amounts of computing power, sophisticated AIs are a reality and they will get better very fast.”[1]

AI ADVANCES AND CHATGPT

In the early 2020s, the vast amount of data available online led to major advances. ChatGPT is one example of this. It is a chatbot created by the company OpenAI. A chatbot writes text responses based on user prompts. These responses may come in many forms. The user might request a

ChatGPT lets users interact with an AI system in the form of a conversation.

simple answer to a question. She may ask the chatbot to write an essay or a fictional story. She may ask it to write a few lines of computer code.

The chatbot creates its answers one word at a time. It uses the data it was trained on to decide what the next word might be. The result is surprisingly humanlike text.

The *GPT* in ChatGPT stands for Generative Pretrained Transformer.

Deep Learning AI

Deep learning AI uses neural networks to process vast amounts of data. Neural networks are algorithms that are designed to process information like the brain does. ChatGPT is a form of deep learning.

This name comes from what it does and how it was made. A transformer is a computer program that quickly figures out how different pieces of data are related. ChatGPT is trained on huge amounts of online text given to it by OpenAI. And it generates responses to user prompts.

Some people are worried about ChatGPT. They think that it could replace jobs in fields such as writing or programming. But that has not happened in a major way yet. That's in part because ChatGPT is not perfect. It makes mistakes. Humans must correct these mistakes. ChatGPT is trained on both real and fictional information from the internet. This is why ChatGPT can write text that is wrong. Sometimes ChatGPT responds with negative **biases**. This is because online information can also have these biases. ChatGPT can even make up false answers. This is called hallucinating.

But with help and guidance from humans, AI can improve how people work and play. ChatGPT and similar chatbots have led to advances in many areas. They are being used to reply to common customer service questions. Writers are using them to come up with ideas. Tutors can use them to create personal tutoring systems. Businesses use them to write basic emails. Medical professionals use them to quickly find information in medical articles. Using AI systems for these kinds of tasks could free up humans to do more meaningful work.

AI systems can help computer programmers write code more efficiently.

WHAT DOES AI LOOK LIKE?

AI can take many forms. ChatGPT is one of many AI systems that make life easier. Examples of AI also include voice assistant products such as Apple's Siri and Amazon's Alexa.

Other forms of AI are embodied. This means they are used in physical things, such as robots or machines. Examples include drones or self-driving cars. Robots in factories also use embodied AI.

Embodied AI includes devices connected to the internet of things (IoT) as well. IoT refers to devices in homes or businesses that can connect to the internet and each other. They collect data using sensors. Cars, thermostats, and kitchen appliances can all be part of the IoT. Applying AI to these devices makes them more capable. For example, self-driving cars have cameras and sensors that see what's around them.

CATEGORIES OF AI

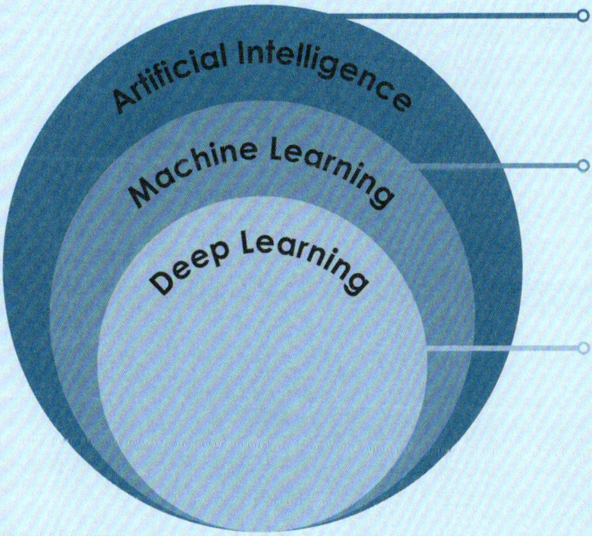

Artificial intelligence involves giving computers the ability to solve problems.

Machine learning involves giving computers the ability to learn from experience and become smarter over time.

Deep learning is a way of doing machine learning that involves a processing structure similar to that of the human brain.

The field of artificial intelligence includes many technical terms. The graphic above shows how some of these terms relate to each other.

AI systems study this data to help detect other cars, pedestrians, or structures along the road. This helps cars drive safely.

CHAPTER TWO

AI AT HOME

AI technology is already making it even easier to relax at home. With AI-driven IoT devices, it's possible to turn on the lights, start the oven, and play music without leaving the couch. This type of AI is called **smart technology**. Homes that use these devices are called smart homes.

In these homes, appliances such as refrigerators and TVs are part of the IoT. So are lights and thermostats. These devices

AI systems can work with IoT devices to help people manage their homes.

17:15
Monday
24°C
Air humidity
5,732
7,146
12,177

Voice-controlled assistants are already common, but AI technology will make them even smarter in the future.

can be controlled through a home automation system. This can be done using an app on a smartphone. But using a virtual assistant such as Google Assistant, Siri, or Alexa works, too. These assistants use AI technology to understand spoken requests and learn a person's preferences. AI is playing a part in many tasks around the house. It is helping to keep homes safe, comfortable, and clean.

The company Josh.ai has combined the IoT with AI. It creates a smart home assistant named Josh. Josh can control many smart devices. It controls fireplaces, appliances, lights, and much more. The company's website explains, "Josh combines all aspects of smart homes into one friendly, well-trained responder. Josh is your companion, and always happy to see you come home."[2]

WHY SMART TECHNOLOGY?

Smart technology can save money. For example, many people forget to turn off lights before leaving home. Smart lighting is one way to solve this problem. Lights can be set to turn on only when people are home. They can also be programmed to turn off when everyone has gone to bed. AI in these devices can learn people's habits and schedules. Then they can adjust settings to save electricity and money.

Smart systems are convenient. Internet access allows people to control these devices from anywhere. For example, people can save time by preheating the

Smart lighting can be adjusted from smartphones and learn people's patterns to set its timing.

oven on the way home from the grocery store. Families on vacation can turn their home's lights on at night. This makes it look like people are home, which might keep burglars away.

Other smart devices make homes more secure. Smart doorbell devices have cameras. They record visitors and package deliveries. Apps let people see who's visiting without going to the door. Microphones and speakers even allow homeowners to talk to those at the door. Video can be reviewed later if a delivered package disappears.

Home security systems may come with motion detectors. Smart locks keep homes safe. They allow entry by fingerprint, voice command, keypad, app, or key. They also keep track of who arrives and leaves through the door. Voice commands can be used to open the door when a person's hands are full.

Smart doorbell devices can use AI to recognize who is at the door.

DOING CHORES

Robots with AI can help people save time on chores. Robots mow the lawn, vacuum carpets, and cook dinner. They can even help clean out the fridge. Some of these robotic assistants have been around for a while. But recent improvements in technology make them even better.

Robotic lawn mowers were first sold in the 1990s. They used an invisible fence system to map out the mower's path. This meant wire needed to be buried to mark the boundaries of the yard. New models are wireless. They use **GPS** tracking to keep the mower on course. Robotic mowers are quiet and mow the lawn at night. The mowers are still not perfect, though. In yards with gardens, these mowers can go off course. Improvements are still being made.

In 2002, Roomba vacuum cleaners became available. These could sense obstacles and change directions. But they often ended up stuck in corners. Robotic vacuums have come a long way. Models built in the early 2020s use AI navigation. They plan cleaning routes based on the layout of a room. Some also mop the floor.

Alexa on Wheels

Amazon Astro is a small, rolling robot with a tablet for a face. It features the company's Alexa voice assistant. Astro can be used for security. It watches over the house and yard. It will send alerts if something isn't right. It plays music and answers questions. It will even fetch snacks.

Robotic vacuum cleaners can use AI to navigate people's homes.

They use sound waves to scrub up dirt. An app allows people to adjust the robot's settings from anywhere.

In the kitchen, robotic cooking arms can help people with disabilities make dinner. They measure and combine ingredients. Kitchen appliances with smart technology can be set to make food or drinks at a certain time. Coffee makers will make a drink in time for breakfast. Refrigerators alert people when food has expired. Some models also suggest recipes based on the ingredients they are storing.

CHAPTER THREE

AI IN ENTERTAINMENT

AI has changed the way people spend their free time. Even basic activities like hanging out with friends have changed. People no longer need to meet at someone's house in order to be together. Now they can chat with friends on Snapchat. Or they can respond to an Instagram post. Social media services use AI to help create feeds of posts, pictures, and videos. Activities like gaming, watching

Social media services use AI to create the feeds that millions of people scroll through each day.

AI can be used to create realistic rival drivers in racing video games.

shows, and listening to music all involve AI as well.

VIDEO GAMES

AI is widely used in video games. It can make virtual worlds and characters seem more real. Companies like Sony are using a form of machine learning to make the gaming experience better. It's called **reinforcement learning**. Reinforcement learning copies the actions of human players. It notices patterns in how people move and do tasks. Then it creates AI-controlled characters using those patterns. Sony CEO Kenichiro Yoshida explains, "By leveraging reinforcement learning, we are developing game AI agents that can be a player's in-game opponent or collaboration partner."[3]

AI can also help create storylines in games. One example is from the game *Middle-earth: Shadow of War*. A company called Monolith created this game. It is based on the *Lord of the Rings series*. *Shadow of War* uses AI for its Nemesis System. This system creates enemy characters with personalities. It uses an algorithm to look for patterns in human behavior. It can remember what happened in earlier battles. Enemy characters improve their skills based on what they learn. This element of surprise makes the game more fun and challenging. Game developer Alexander Goldybin explains, "One of the key benefits of using AI in video games

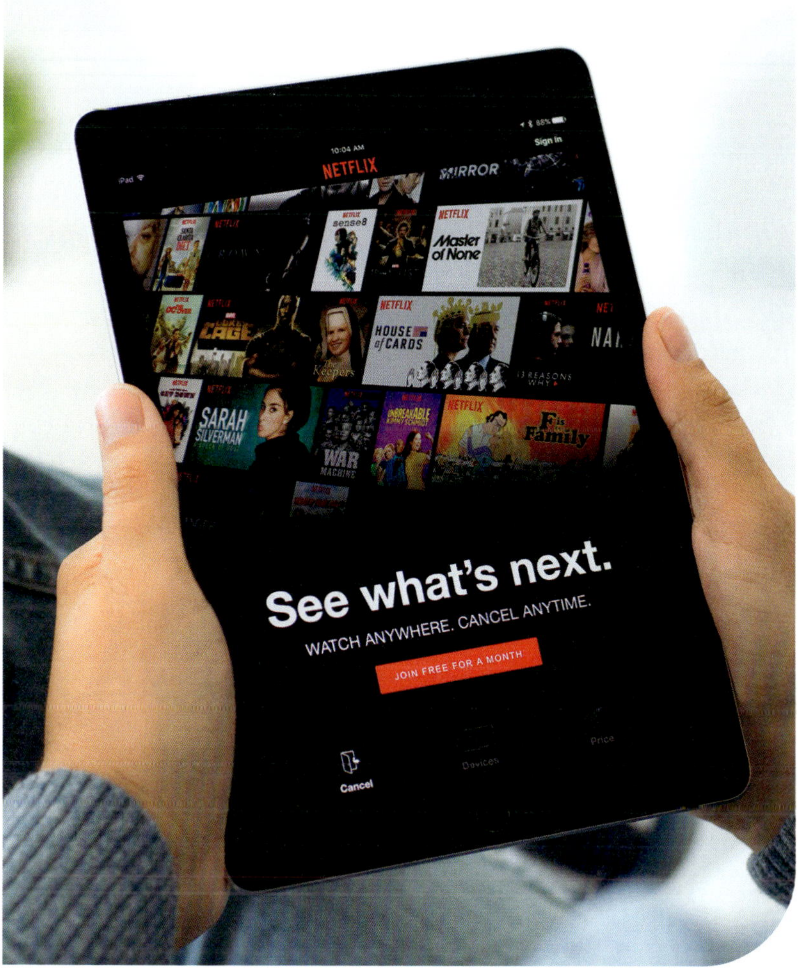

Services such as Netflix use AI to predict what shows and movies a viewer will like.

is that it allows for more dynamic and responsive gameplay."[4]

STREAMING SERVICES

AI makes it easier for people to find a movie or show they like. Streaming services such as Netflix offer many options. Sometimes it can be hard to choose what to watch. AI makes this easier. Deep learning algorithms see patterns in what viewers watch. These algorithms look for aspects such as genre or topic. They also look for favorite actors or directors. The algorithms use these patterns to make lists of shows that viewers might enjoy.

AI and Shopping

Online retailers such as Amazon use AI to recommend products to customers. AI also helps monitor demand for specific items. This helps retailers keep products in stock. Some clothing and accessory brands even have a virtual try-on feature. It uses generative AI to show how products might look on a range of body types.

Music streaming services such as Spotify work the same way. Genres, artists, and listening history shape the listener's profile. In 2023 Spotify began its DJ program. **Generative AI** uses the patterns in gathered data to make the ideal playlist for each listener. The more a person uses the service, the better the playlist becomes.

SOCIAL MEDIA

Like streaming services, social media platforms collect data with algorithms. Facebook uses AI to track the types of posts users comment on. It also notes who users follow and interact with most. Based on this data, AI suggests content and people to follow. Instagram uses AI in the same way.

TikTok is a popular social media platform that uses AI.

Snapchat is known for its filters. They use AI facial recognition to study people's photos. The filters can then make people look like paintings, anime characters, and more. AI also tracks users' interests to deliver personalized ads.

TikTok also heavily uses AI. In August 2023, the video service had 1.67 billion users. TikTok tracks the content people watch. It also monitors how they react to content. It uses this to show people more videos they might like.

CHAPTER FOUR

AI AT SCHOOL

The rise of advanced AI chatbots in the early 2020s brought many new questions for schools. Teachers and parents worried that students would use services such as ChatGPT to cheat. Schools struggled with what to do about this new technology.

The question of how to handle ChatGPT in the classroom was new. But the use of AI in schools wasn't. For many years,

In the early 2020s, teachers, students, and parents worked to figure out the role of advanced AI technology in classrooms.

teachers have used simple AI to grade tests and complete other repetitive tasks. AI systems have provided personalized learning. These systems meet students where they are. There are also learning apps that can help students with any subject, at any time of day. These are just a few ways AI is being used in the classroom.

AI FOR TEACHERS

According to a 2019 survey, teachers spend 57 percent of their time on non-teaching tasks. These include grading, planning lessons, and doing paperwork. AI can speed up these tasks. This helps teachers focus on connecting with students. AI also helps them research and find creative ways to teach concepts. Simple AI systems have

AI could make grading homework easier and more efficient for teachers.

been used to grade multiple-choice tests for decades. Newer vision-based AI allows computers to grade written tests. Teachers take a photo of a student's math homework. Then they upload it to a software program for grading.

Machine learning AI has also been used to grade complicated work. Software such as Gradescope and E-rater

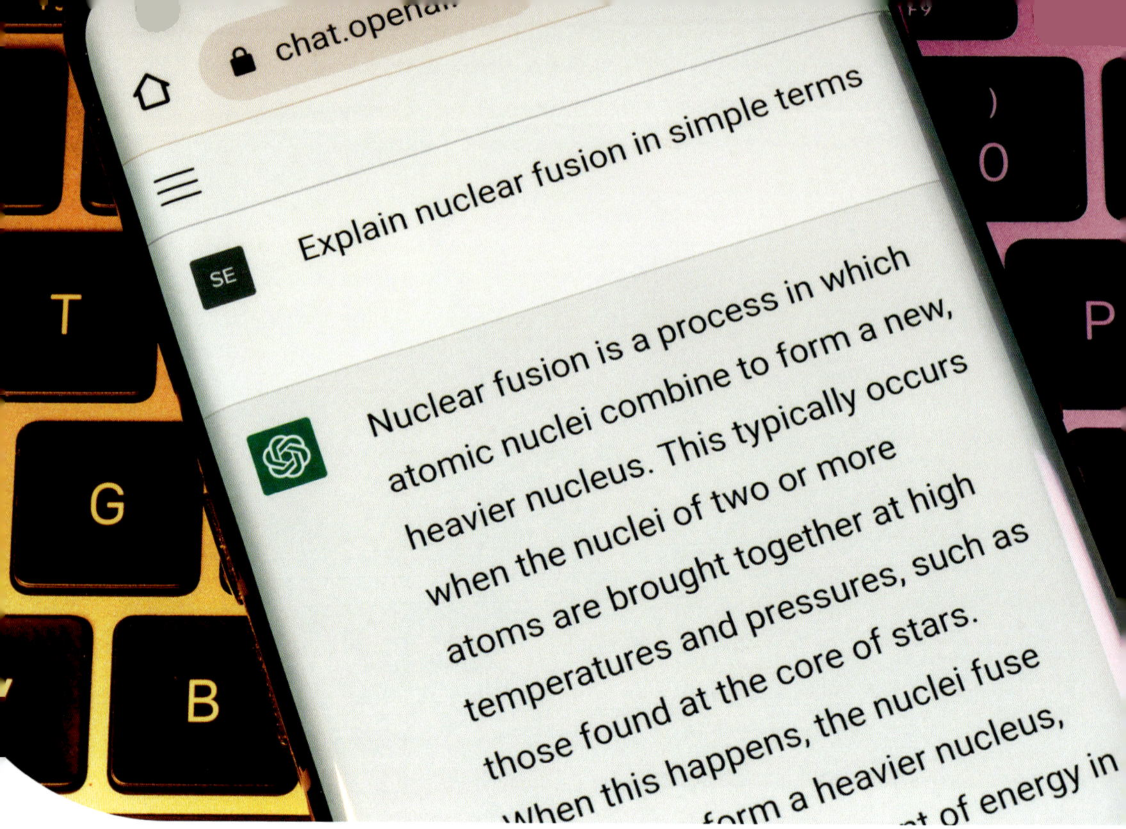

Students and teachers have begun learning the strengths and weaknesses of AI products such as ChatGPT.

use ML to evaluate written responses and essays. These tools look at vocabulary and organization. They also review how sentences are structured.

AI is also being used to catch gaps in learning. Tools using AI alert the teacher when a student gives a wrong answer or misses questions about a specific topic.

The teacher can then help the student learn the concept. Administrators are using something similar. It's called an early warning system. Nearly 50 percent of high schools and 90 percent of colleges use it to keep track of student attendance and grades. When these dip, administrators and teachers reach out to help the struggling students.

CHATGPT IN THE CLASSROOM

Instead of trying to stop students from using ChatGPT, some teachers are embracing the tool. They are using it to help students learn. Kevin Roose is a technology writer for the *New York Times*. He has written about emerging issues with AI systems. Roose explains, "Today's students will graduate into a world full of generative [AI] programs. They'll need to know their way around these tools—their strengths and weaknesses, their hallmarks and blind spots—in order to work alongside them."[5]

Teachers have come up with many ways to use ChatGPT. Some teachers have students use it to write outlines for essays. Then students put away their laptops and write the essay themselves. This gets students to think critically and develop editing skills. Nicole Haddad is an instructor at Southern Methodist University. She explains, "Before they even use ChatGPT, I help students discern what is worth knowing, figuring out how to look it up, and what information or research is worth 'outsourcing' to [AI]."[6]

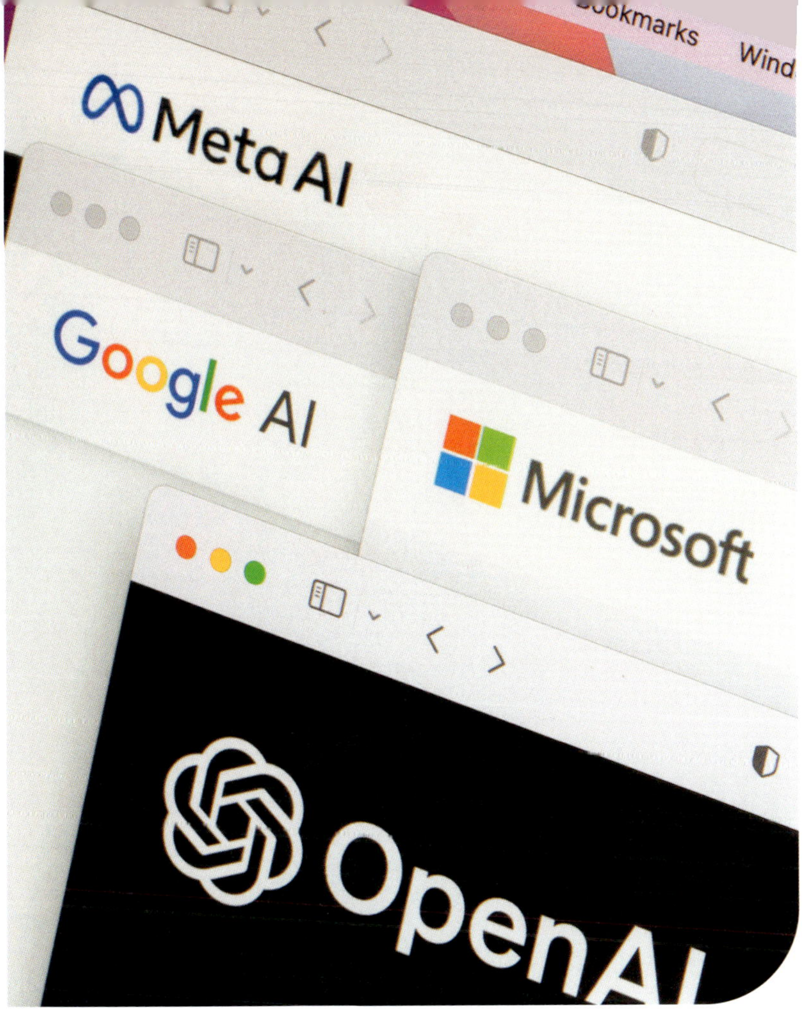

Meta, Google, and OpenAI are among the major companies working on generative AI systems.

One teacher asked students to use ChatGPT to write rap lyrics about similes and metaphors. Then the class edited the rap and added a beat. Teachers have been using ChatGPT to dream up new quiz questions. And they use it to make sure

AI-powered tutoring services could open up personalized learning to more students.

instructions are clear. They also prompt the chatbot for assignment ideas and discussion topics.

KHANMINGO

Khanmingo is a tutoring chatbot. It gives personal instruction to students. It uses OpenAI's GPT-4 technology. Khanmingo teaches at a higher level than earlier tutoring programs. It doesn't just correct wrong answers. The program also asks students to explain their thinking.

Many people seem to like it. Zaya is a sixth grader in Palo Alto, California. She said, "It's very good at walking you through the problem step by step. Then it congratulates you every time it helps you solve a problem."[7]

There's an App for That

Students use many AI-driven apps to do schoolwork. Grammarly helps with writing. Duolingo provides language practice. Socratic provides visual explanations of math, science, and language arts topics. Tutor a.i. allows students to create their own tutoring avatar to help with science, math, history, or literature. Copyscape makes it easy to check text for plagiarism.

Khanmingo offers tutoring in history, civics, and chemistry. The chatbot even allows students to have virtual conversations with historical figures such as Marie Curie. There are downsides, of course. Khanmingo isn't always right, so students and teachers need to be on alert for wrong answers.

THE FUTURE OF AI

As AI continues to improve, it will be used in more areas of everyday life. Just as robotic vacuums and lawnmowers have advanced, existing AI technologies will be refined. Robots will do more chores around the house. Wearable devices will monitor people's health. Autonomous cars will communicate with each other directly.

As AI changes over time, so will its impact on people's daily lives.

This will allow them to navigate and avoid accidents. Scientists will continue to look for other ways that AI can be used to improve daily life. The possibilities are endless.

GLOSSARY

biases

existing feelings someone has about someone or something

generative AI

AI that uses patterns in gathered data to create new content

GPS

global positioning system, which uses signals from satellites to determine a device's location

machine learning (ML)

AI that allows systems to learn and improve from experience

reinforcement learning

an area of machine learning that allows programs to learn from mistakes

smart technology

a term referring to devices controlled by computers or AI

SOURCE NOTES

CHAPTER ONE: AI AND HOW IT WORKS

1. Quoted in Ayush Jalan, "10 Realistic Predictions for the Next 30 Years that Will Change How We Live," *Make Use Of*, April 8, 2023. www.makeuseof.com.

CHAPTER TWO: AI AT HOME

2. Quoted in "Pioneering a New Era of Intelligent Living," *Josh.ai*, 2024. www.josh.ai.

CHAPTER THREE: AI IN ENTERTAINMENT

3. Quoted in Keith Stuart, "Think, Fight, Feel: How Video Game Artificial Intelligence Is Evolving," *The Guardian*, July 19, 2021. www.theguardian.com.

4. Quoted in Alexander Goldybin, "Using AI Games in Development: Part Two," *Pocket Gamer.biz*, June 30, 2023. www.pocketgamer.biz.

CHAPTER FOUR: AI AT SCHOOL

5. Quoted in Kevin Roose, "Don't Ban ChatGPT in Schools. Teach With It," *New York Times*, January 12, 2023. www.nytimes.com.

6. Quoted in Matt Miller, "ChatGPT, Chatbots and Artificial Intelligence in Education," *Ditch That Textbook*, December 17, 2022. www.ditchthattextbook.com.

7. Quoted in Natasha Singer, "New A.I. Chatbot Tutors Could Upend Student Learning," *New York Times*, June 8, 2023. www.nytimes.com.

FOR FURTHER RESEARCH

BOOKS

Jennifer Kaul, *The Potential of Artificial Intelligence*. San Diego, CA: BrightPoint Press, 2025.

Pamela McCauley, *Engineering for Teens*. New York: Rockridge Press, 2021.

Dr. Claire Quigley, Simply Artificial Intelligence. New York: DK, 2023.

INTERNET SOURCES

Alex Hughes, "ChatGPT: Everything You Need to Know About OpenAI's GPT-4 Tool," *BBC Science Focus*, September 25, 2023. www.sciencefocus.com.

Will McCurdy, "No More 'I Took an Arrow to the Knee': Could AI Write Super-intelligent Video Game Characters?" *Guardian*, May 25, 2023. www.theguardian.com.

Emily A. Vogels, "A Majority of Americans Have Heard of ChatGPT, but Few Have Tried It Themselves," *Pew Research Center*, May 24, 2023. www.pewresearch.org.

WEBSITES

OpenAI
www.openai.com

OpenAI is the company that developed ChatGPT. Its website explains the uses for OpenAI products and provides information about types of AI. The site also includes articles about OpenAI research.

What Is Artificial Intelligence—And How Does It Affect You?
www.voicesofyouth.org/ai

Voices of Youth is a website by UNICEF that provides information about AI to kids and teens. It answers questions about AI and explains AI rights. The site also includes videos of kids giving their opinions about AI.

Why AI
www.ai.google

Why AI is a website run by Google on making AI accessible to everyone. It includes articles about how AI is being used responsibly to help people and society.

INDEX

Alexa, 22, 27, 33
algorithm, 15, 18, 40, 42, 43
Amazon, 22, 42
Amazon Astro, 33
AI wearables, 6, 8–9
autonomous cars, 9, 56

ChatGPT, 17–21, 22, 46, 51–53

deep learning AI, 18, 23, 42
drones, 22

face recognition systems, 12
Facebook, 43

Gates, Bill, 16
generative AI, 42, 43
GPS, 32
Goldybin, Alexander, 40–41
Google Assistant, 27

Haddad, Nicole, 52
Instagram, 36, 43
internet of things (IoT), 22–23, 24, 27

Josh.ai, 27–28

Khanmingo, 54–56

Lord of the Rings, 40

machine learning, 15, 16, 23, 39–40, 49
Microsoft, 16

Nemesis System, 40
Netflix, 12, 42

OpenAI, 17–19, 54

reinforcement learning, 39
robotic cooking arms, 35
robotic lawn mower, 32
robotic vacuums, 33–35, 56
robots, 22, 32, 56
Roomba, 33

self-driving cars, 22, 23
Siri, 22, 27
smart technology, 24, 28–30, 35
Snapchat, 36, 45
Sony, 39
Spotify, 43

teachers, 46, 48–53, 56
TikTok, 45
Tutor a.i., 55

wearable devices, 56

Yoshida, Kenichiro, 39–40

IMAGE CREDITS

Cover: © kenchiro168/Shutterstock Images
5: © Panuwat Phimpha/Shutterstock Images
7: © Vlasov Yevhenii/Shutterstock Images
8: © melamorworks/Shutterstock Images
11: © jamesteohart/Shutterstock Images
13: © New Africa/Shutterstock Images
14: © NicoElNino/Shutterstock Images
17: © Ascannio/Shutterstock Images
21: © Gorodenkoff/Shutterstock Images
23: © Red Line Editorial
25: © Gorodenkoff/Shutterstock Images
26: © r.classen/Shutterstock Images
29: © RossHelen/Shutterstock Images
31: © voronaman/Shutterstock Images
34: © labelled/Shutterstock Images
37: © Rawpixel.com/Shutterstock Images
38: © Tinxi/Shutterstock Images
41: © DenPhotos/Shutterstock Images
44: © BigTunaOnline/Shutterstock Images
47: © goodluz/Shutterstock Images
49: © Monkey Business Images/Shutterstock Images
50: © Ascannio/Shutterstock Images
53: © Tada Images/Shutterstock Images
54: © Ground Picture/Shutterstock Images
57: © Somkid Thongdee/Shutterstock Images

ABOUT THE AUTHOR

Kari Cornell is an award-winning children's book author who gardens, runs, and makes pottery. She lives in Minneapolis with her husband, two boys, and their sweet dog, EmmyLou.